INTERNATIONAL CENTRE FOR MECHANICAL SCIENCES

COURSES AND LECTURES - No. 46

JOHN D. CAMPBELL

UNIVERSITY OF OXFORD

DYNAMIC PLASTICITY OF METALS

COURSE HELD AT THE DEPARTMENT
FOR MECHANICS OF DEFORMABLE BODIES
JULY 1970

UDINE 1970

SPRINGER-VERLAG WIEN GMBH

ISBN 978-3-211-81149-8   ISBN 978-3-7091-2848-0 (eBook)

DOI 10.1007/978-3-7091-2848-0

PREFACE

The study of the dynamic plasticity of metals has become increasingly important in recent years for two reasons. Firstly, it has become clear that many practical problems such as those associated with brittle fracture, wave propagation and high rate forming can only be satisfactorily treated if the rate dependence of the flow behaviour of the material is taken into account. Secondly, the development of dislocation theory has shown that the micro-mechanisms governing plastic flow are usually essentially rate sensitive, and hence a study of rate dependence is a powerful tool in elucidating the basic flow processes.

In the present monograph, which is the text of my lectures given at CISM in October 1970, I have tried to summarize the progress which has been made in the past two decades in understanding both macroscopic and microscopic dynamic flow behaviour. Much of the difficulty of the subject lies in the problem of devising reliable experimental methods, particularly at very high rates of strain. A considerable part of the contents is therefore devoted to a discussion of this problem and an account of available experimental techniques. Clearly, it has not been possible to include more than a very limited number of experimental results, but I hope that I have given a representative selection indicating the main types of observed *behaviour.*

I wish to express my appreciation of the privilege of taking part in the work of CISM, and in particular to thank Professors Sobrero and Olszak for inviting me to give this course of lectures.

J. Deans Campbell

Udine, July 1970

Chapter 1.

Plastic Deformation of Metals.

1.1. Phenomenological Theories.

The classical theory of plasticity is based on the concept of a <u>yield criterion</u> $f(\sigma_{ij}) = k$, where σ_{ij} is the stress tensor and k is a material constant, which is in general a function of the previous strain history. For an isotropic material, the function f may be expressed in terms of the three invariants I_1, I_2, I_3, of the stress tensor. Experimentally, it is found that the hydrostatic component of stress does not affect plastic flow, so that the stress may be replaced by the <u>stress deviation</u>

$$S_{ij} = \sigma_{ij} - \frac{1}{3}\sigma\delta_{ij} , \qquad (1.1.1)$$

where $\sigma = \sigma_{kk}$ and δ_{ij} is the Kronecker delta. The yield criterion thus reduces to $f(J_2, J_3) = k$, where J_2, J_3 are the second and third invariants of S_{ij} (J_1 being identically zero). In the von Mises criterion this is further simplified to $J_2^{1/2} = k$ where k is now identified with the yield stress in simple shear.

For a material which work-hardens isotropically, k may be expressed as a function of the work of plastic deformation

$$W = \int \sigma_{ij}\, d\varepsilon_{ij}^p , \qquad (1.1.2)$$

where ε_{ij}^{p} is the plastic component of strain. For a non-work-hardening or ideally plastic material, k is a constant.

The plastic strain-rate $\dot{\varepsilon}_{ij}^{p}$ is governed by the flow rule, according to which

$$(1.1.3) \qquad \dot{\varepsilon}_{ij}^{p} = \lambda \frac{\partial f}{\partial \sigma_{ij}} ,$$

where λ is a scalar factor of proportionality. This rule determines only the relative plastic strain rates; the stress-strain relation is independent of time.

The total strain rate is given by adding the elastic component to the plastic component:

$$(1.1.4) \qquad \dot{\varepsilon}_{ij} = \frac{1}{2G} \dot{s}_{ij} + \frac{1 - 2\nu}{E} \dot{\sigma}_m \delta_{ij} + \lambda \frac{\partial f}{\partial \sigma_{ij}} ,$$

where $\sigma_m = \sigma_{kk}/3$, G is the shear modulus, E is Young's modulus, and ν is Poisson's ratio.

To generalize (1.1.4) for rate-dependent materials, the concept of a visco-plastic material may be introduced. Such a material behaves elastically for stresses such that $f(\sigma_{ij}) < k$ but requires that $f(\sigma_{ij}) > k$ for finite rates of plastic strain. Following Sokolovsky [1] and Malvern [2] , it may be assumed that the plastic strain rate is a function of the overstress, i.e. the amount by which the applied stress exceeds that corresponding to the flow stress for the same strain at vanishingly small strain rates.

Perzyna [3] has proposed the following constitutive law based on this concept:

$$\dot{\varepsilon}_{ij} = \frac{1}{2G}\dot{s}_{ij} + \frac{1-2\nu}{E}\dot{\sigma}_m\delta_{ij} + 2\gamma\langle\Phi(F)\rangle\frac{\partial f}{\partial\sigma_{ij}}, \quad (1.1.5)$$

where $F = f/k - 1$, γ is a material constant, Φ is an arbitrary function, and the symbol $\langle\;\rangle$ is defined thus:

$$\langle x \rangle = \begin{cases} 0 & \text{for} \quad x \leq 0 \\[2mm] x & \text{for} \quad x > 0 \end{cases} \quad (1.1.6)$$

According to (1.1.5), the plastic strain rate is independent of the strain history, depending only on the instantaneous values of stress and strain; (1.1.5) is therefore a "mechanical equation of state" relating stress, strain and their time derivatives.

Adopting the Mises yield criterion, $f = J_2^{1/2} = (\frac{1}{2}s_{ij}s_{ij})^{1/2}$ and $\partial f/\partial\sigma_{ij} = \partial f/\partial s_{ij} = \frac{1}{2}s_{ij}/J_2^{1/2}$, so (1.1.5) becomes

$$\left.\begin{aligned} \dot{\varepsilon}_{ij} &= \frac{1}{2G}\dot{s}_{ij} + \gamma\langle\Phi(\frac{J_2^{1/2}}{k}-1)\rangle\frac{s_{ij}}{J_2^{1/2}} \quad (i\neq j) \\[2mm] \dot{\varepsilon}_{ii} &= \dot{\sigma}_m/K, \end{aligned}\right\} \quad (1.1.7)$$

and

where K is the bulk modulus.

For a test under uniaxial tension σ , (1.1.7) becomes

$$\dot{\varepsilon} = \frac{\dot{\sigma}}{E} + \frac{2\gamma}{\sqrt{3}}\langle\Phi(\frac{\sigma}{\sigma_0}-1)\rangle, \quad (1.1.8)$$

where ε is the strain and $\sigma_0 = \sqrt{3}\,k$ the uniaxial tensile yield stress. The theory thus predicts that at constant plastic strain

rate the stress at a given strain will be a multiple of the static yield stress. It follows that the observed work-hardening rate
$d\sigma/d\varepsilon$ will increase with increasing strain rate. An alternative
assumption to that made in (1.1.5) would be to take $F = (f - k)/C$,
where C is a constant. In this case the theory would predict that
the observed work-hardening rate would remain constant at any
plastic strain rate. In a more general formulation, f is taken to
depend on the plastic strain as well as the stress [3] .

The effect of temperature may be allowed for by
assuming that the quantities γ and k, and possibly the function
Φ in (1.1.7) are functions of the temperature T. Perzyna [3] has
shown that it is possible to correlate several series of experimental data for mild steel, pure iron and aluminium by taking γ
and k as functions of T, while the function Φ is independent of T.
Lindholm [4] has presented experimental data obtained in combined
tension and shear tests on aluminium. He has shown that they are
in fair agreement with (1.1.5) when γ and Φ are temperature dependent, the particular functions of temperature being based upon a
dislocation model for plastic flow. It was, however, necessary
to use the generalized form $f = f(\sigma_{ij}, \varepsilon^p_{kl})$ to obtain this agreement (see Section 1.2).

Equation (1.1.8) is a special case of the general
relation

(1.1.9) $\dot{\varepsilon}^p = h(\sigma, \varepsilon^p)$.

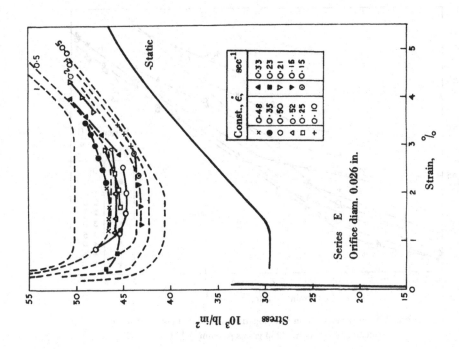

Fig. 2. Stress/strain curves derived from damped dynamic tests
(mean grain density, 2033 grains per mm²). [5]

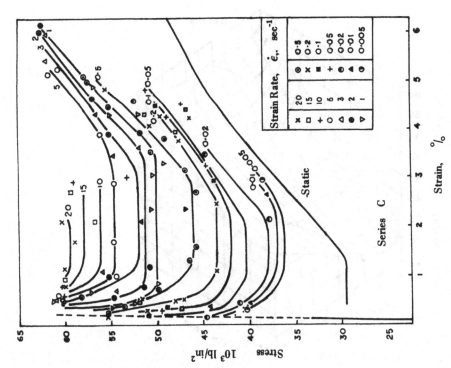

Fig. 1. Stress/strain curves derived from constant-stress tests
(mean grain density, 2033 grains per mm²) [5]

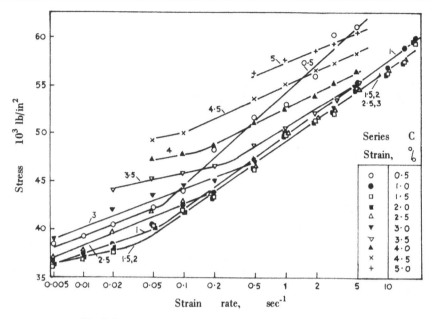

Fig. 3. Stress against strain rate (logarithmic scale) at constant strain
(mean grain density, 2033 grains per mm^2). [5]

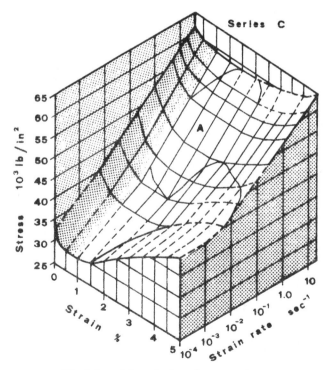

Fig. 4. Isometric projection of stress, strain, strain-rate surface
(mean grain density, 2033 grains per mm^2). [5]

Equations of the form (1.1.9) have been postulated by many investigators, and this form is also indicated by dislocation theory
if strain history effects can be neglected. It has been shown [5]
that such a relation is valid, to a first approximation, for mild
steel at moderate strains and strain rates (Figs.1–4, see pp.9,
10). However, recent work on other bodycentred cubic metals [6]
has indicated that considerable deviations may occur when the
strain rate is suddenly altered during the course of a test;
strain–history effects have also been observed in tests on aluminium [7] . Such effects are considered in Section 3.1.

Many experimental investigations are performed on
simple tension specimens and the possibility must therefore be
considered that the deformation will become unstable, so that
necking of the test piece occurs. The problem of such instability has been analysed on the assumption that the material obeys
equation (1.1.9) [8] , the elastic strains being negligible. It is
found that the strain gradients along the specimen increase indefinitely with time if

$$\frac{\partial h}{\partial \varepsilon^p} + \frac{\sigma}{1 + \varepsilon^p} \frac{\partial h}{\partial \sigma} < \frac{\dot{\varepsilon}^p}{1 + \varepsilon^p} . \qquad (1.1.10)$$

For a material with a limiting "static" stress-
strain curve $h(\sigma, \varepsilon^p) = 0$, (1.1.10) gives the same condition as Con-
sidère's construction. In general, (1.1.10) is satisfied in some
region of the (σ, ε^p) plane. For materials such as mild steel,
this region is defined by a curve with negative slope $\partial \sigma / \partial \varepsilon^p$;

thus the necking strain, at which instability develops, decreases
as the strain rate is increased (Fig.5).

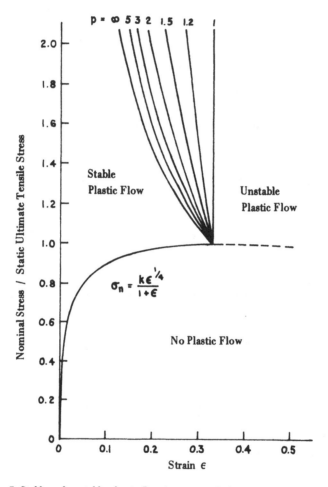

Fig. 5. Stable and unstable plastic flow in a material obeying the constitutive relation
$$\dot\epsilon = C\,(\sigma - k\epsilon^{1/4})\,p. \quad [8]$$

This is shown by experimental results [9] obtained at strain
rates up to $106\ \mathrm{s}^{-1}$ (Fig.6). The effect of adiabatic heating

on the stability criterion has been discussed by Klepaczko [10].

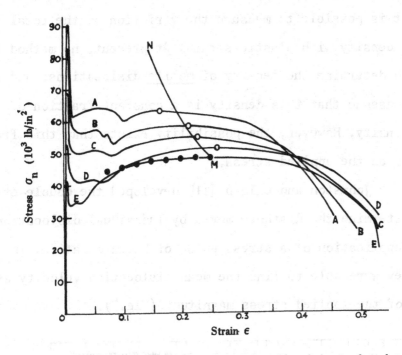

Fig. 6. Experimental dynamic stress-strain curves for mild steel, showing derived stability
boundary MN. [8]

1.2. Dislocation Theory.

It is known that in metals the predominant mechanism of plastic flow is that of slip caused by the motion of dislocations. The basic kinematic relation between the plastic shear
strain rate $\dot{\gamma}^p$, the density of mobile dislocations ρ_m and their
mean velocity v is

$$\dot{\gamma}^p = b\,\rho_m v \qquad\qquad (1.2.1)$$

where b is the magnitude of the Burgers vector.

By means of etch – pit techniques or electron mi-
croscopy, it is possible to measure the variation of the total
dislocation density with plastic strain. At present, no method is
available to determine the density of <u>mobile</u> dislocations, and it
is usual to assume that this density is a constant fraction of
the total density. However, the possibility exists that this frac
tion depends on the applied stress.

Johnston and Gilman [11] developed the double-etch
method to determine the distance moved by individual dislocations
during the application of a stress pulse of known duration. In
this way they were able to find the mean dislocation velocity as
a function of the applied stress magnitude (Fig.7).

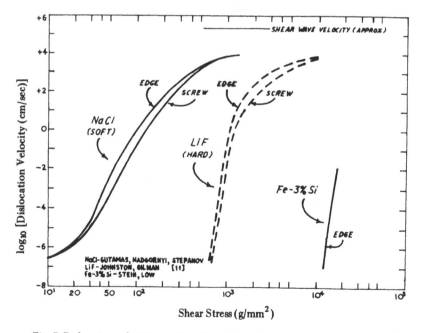

Fig. 7. Dislocation velocities as a function of stress for some representative crystals at
room temperature.

Such a relation is only valid if the effective inertia of a dislocation is negligibly small, so that the time required to accelerate it is zero. An estimate of this acceleration time for aluminium [12] gives a value of 10^{-9} s, which is several orders of
magnitude smaller than the times involved even in high-speed impact experiments. Thus it may be assumed that v is a function of
σ, the applied stress. Equation (1.2.1) therefore takes the form
(1.1.9), provided that ρ_m does not depend on strain history. Elec-
tron-microscope studies of niobium subjected to slow and impact
compression [13] show that the relation between the total dislocation density ρ and the plastic strain is independent of
strain rate; however, the distribution of dislocations is found
to be much more uniform at high rates, so that some variation
in ρ_m with strain rate is possible.

An alternative expression for the plastic strain
rate due to dislocation motion may be obtained by considering a
model due to Seeger [14]. In this model, dislocations are held
up at N points per unit volume, where they interact with other
dislocations. Plastic flow occurs at a rate governed by the frequency v with which dislocations can intersect each other, and
the area A swept out by a dislocation after such an intersection
has occurred. The plastic strain rate is thus given by

$$\dot{\gamma}^{p} = NAb\nu .\qquad\qquad (1.2.2)$$

If the applied stress is not great enough to enable intersection to occur, the frequency ν will be governed by thermal activation. If G is the free energy of activation, we may write

$$(1.2.3) \qquad \nu = \nu_0 \exp\left(-G/kT\right),$$

where ν_0 is the frequency with which intersection attempts are made, k is Boltzmann's constant, and T is the absolute temperature.

The activation energy G is reduced by the applied stress, since work is done by this stress during the intersection process. If we assume that this work varies linearly with the applied stress, we have

$$(1.2.4) \qquad G = G_0 - (\tau - \tau_i)V ,$$

where τ_i is an effective internal stress due to long-range stress fields, V is a quantity of the dimensions of volume (activation volume), and G_0 is the total activation energy for intersection.

Combining (1.2.2), (1.2.3) and (1.2.4) we obtain

$$(1.2.5) \qquad \dot{\gamma}^p = NAb\nu_0 \exp\left(\frac{-G_0}{kT}\right)\exp\left[\frac{(\tau - \tau_i)V}{kT}\right].$$

The activation volume V is a function of the plastic strain, since it depends on the mean dislocation network size. Thus (1.2.5) implies that in (1.1.5) the yield function F and hence f, depends on both stress and plastic strain.

Lindholm [4] has used a generalized form of (1.2.5)

to correlate data obtained in tests on aluminium under various
stress conditions, at strain rates up to $10^3 s^{-1}$ (Figs. 8,9,10).
He has also carried out some tests under non–proportional loading, which indicated that the effect of loading history is not
large.

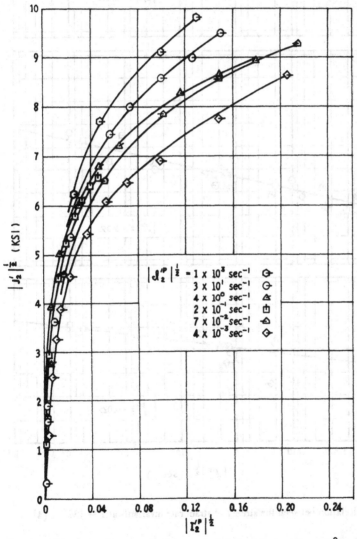

Fig. 8. Relation between the stress and strain invariants at T = 294° K. [4]

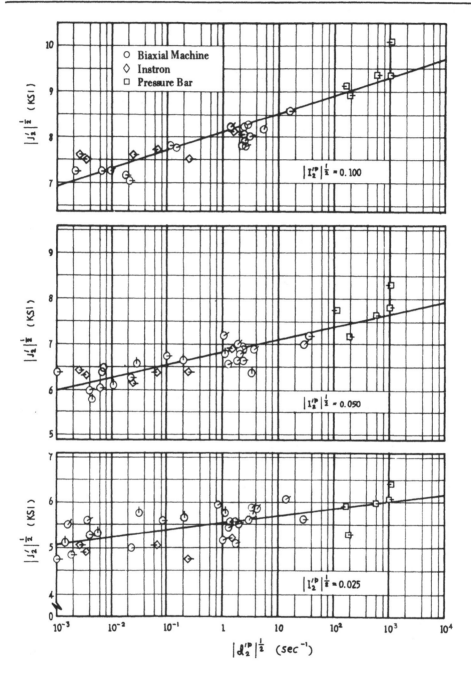

Fig. 9. Relation between the stress and strain-rate invariants at T = 294° K. [4]

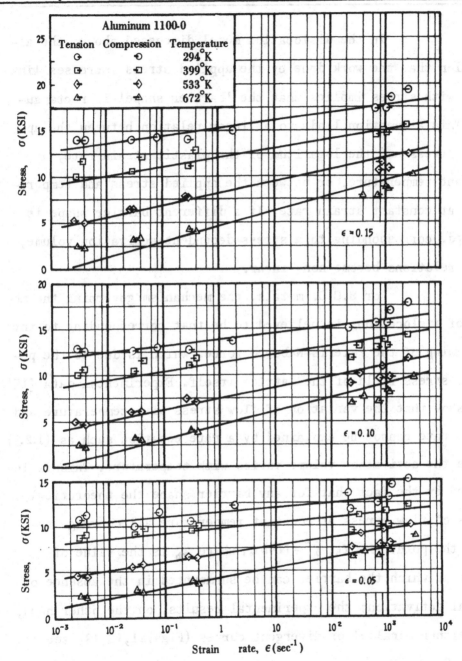

Fig. 10. Stress vs. strain-rate at constant temperature and strain. [4]

In the dislocation model discussed above, the assumption that the work done by the applied stress increases linearly with stress implies that the 'barrier shape' is rectangu-
lar. This assumption leads to a linear relation between the applied stress and the logarithm of the plastic strain rate, at
constant temperature, or between the applied stress and temperature, at constant strain rate. If a different barrier shape is
assumed, corresponding to a stress-dependent activation volume,
these relations become non-linear.

For B.C.C. metals, the mechanism governing the motion of dislocations is believed to be that corresponding to the
overcoming of the Peierls-Nabarro force, which is due to the pe-
riodic stress field of the lattice itself. Experimental data [15],
[16] show that the variation of flow stress with temperature and
strain rate cannot be explained by a rate equation such as (1.2.5)
unless the activation volume varies with temperature. This is indicated by the fact that for any barrier shape the theoretical
curves of flow stress at constant temperature all converge to-
wards the point $\tau = \tau_b$, $\dot{\gamma}^P = NAb\nu_0$, where τ_b is the value of
stress at which the barrier can be surmounted in the absence of
thermal activation; the experimental results, on the other hand,
show either parallel or divergent curves (Figs.11,12,13, see pp.
21 and 22).

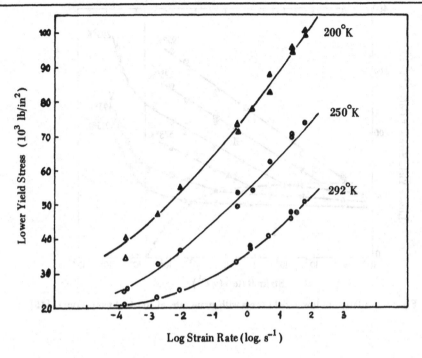

Fig. 11. Variation of the Lower Yield Stress of Niobium with the logarithm of the Strain Rate. [6]

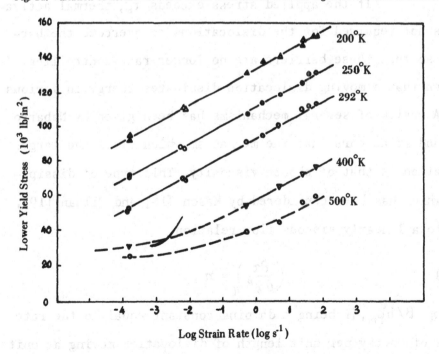

Fig. 12. Variation of the Lower Yield Stress with the logarithm of the Strain Rate for Molybdenum. [6]

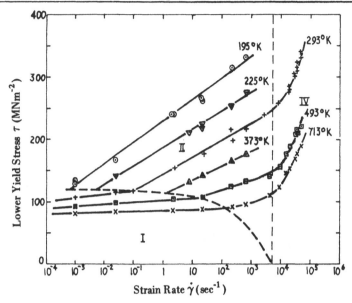

Fig. 13. Variation of lower yield stress with strain rate, at constant temperature. [15]

If the applied stress exceeds τ_b, thermal activation is not required for the dislocations to overcome the barriers, so that these barriers are no longer rate-controlling. It is known that a moving dislocation dissipates energy in various ways. A review of several mechanisms has been given by Nabarro [17], and it appears that the mechanism which gives the largest dissipation is that of phonon viscosity. This type of dissipation, which has been considered by Mason [18] and Gilman [19], leads to a linearly viscous flow relation

$$(1.2.6) \qquad \left(\frac{\partial \tau}{\partial \dot{\gamma}^P}\right)_T = \eta ,$$

where $\eta = B / b^2 \rho_m$, B being a damping constant equal to the rate of loss of energy per unit length of dislocation moving at unit velocity.

For applied stresses exceeding the barrier stress τ_b therefore, the flow stress is given by

$$\tau = \tau_b + \eta \dot\gamma^P, \tag{1.2.7}$$

where

$$\tau_b = \tau_i + G_0/V \tag{1.2.8}$$

from (1.2.4).

Experiments on zinc single crystals [20] and aluminium single crystals [21] show that (1.2.7) is obeyed at strain rates exceeding about $10^3 s^{-1}$ (Figs.14,15, see below and p.24). For both metals, τ_b was found to be independent of temperature; for zinc, η was also sensibly independent of temperature, while for aluminium η decreased as the temperature increased from 20°

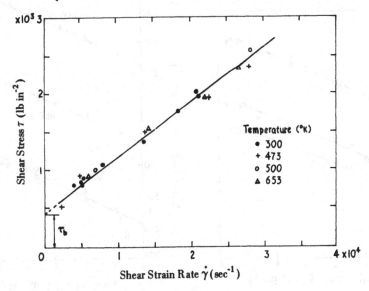

Fig. 14. Shear stress plotted against shear strain rate for basal shear in zinc single crystals.[20]

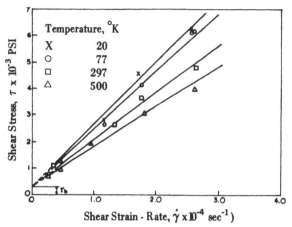

Fig. 15. Shear stress vs. shear strain rate for dynamic shear in alluminium single crystals.[21]

to 500°K. Experiments on polycrystalline low-carbon steel [15] also showed agreement with (1.2.7) at strain rates above about $5 \times 10^3 \text{ s}^{-1}$; for this material, however, τ_b showed a considerable variation with temperature, while η varied little in the range 293 to 713°K (Fig. 16). The decrease of τ_b with increasing temperature in this range is ascribed to an increase of V with T.

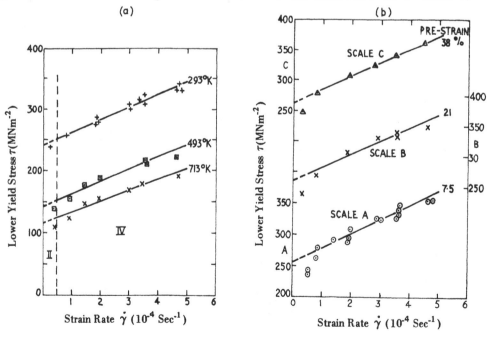

Fig. 16. Variation of lower yield stress with strain rate (region IV). (a) Zero pre-strain; temperature 293, 493, 713°K. (b) Pre-strain 7·5, 21, 38°$_b$; temperature 293°K. [15]

The above discussion is based on the assumption that the measured stress and deformation is representative of that throughout the gauge length of the specimen. However, in certain cases the plastic strain may become heterogeneous during the early stages of the deformation. This may be detected by surface markings, or by sectioning and etching (Fig. 17). If it is

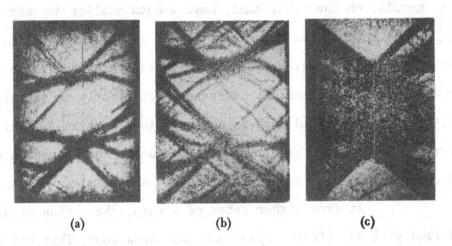

(a)   (b)   (c)

Fig. 17. Macrophotographs showing non-uniform yield in compression specimens deformed at constant stress. Longitudinal sections through axes of specimens $^3/_8$ in. diam., $^1/_2$ in. long. Amount of deformation increases from (a) to (c).

assumed that the mobile dislocation density ρ_m is a function of plastic strain only, and the mean dislocation velocity $\mathbf{v}$ is a function of the stress only, it may be shown [9] that the gradient of the dislocation density increases exponentially with time if

$$\frac{d}{d\gamma^p}\left[\ln\left(\rho_m\frac{d\rho_m}{d\gamma^p}\right)\right] > G\frac{d}{d\tau}(\ln\mathbf{v}). \qquad (1.2.9)$$

The inequality (1.2.9) is satisfied for certain

values of stress and plastic strain, and if such values can be
reached the material will show plastic heterogeneity, which is
often associated with a yield drop. Mild steel shows this effect
at ordinary rates of strain, as the density of mobile dislocations is initially low, due to pinning by carbon and nitrogen
atoms, and increases rapidly with increasing plastic strain.
F.C.C. metals, on the other hand, have a much smaller dependence
of ρ_m on γ^p so that the inequality is not satisfied at ordinary
rates. It appears, however, that at very high rates the value of
$d(\ln v)/d\tau$ for such materials decreases sufficiently to satisfy
(1.2.9), so that a yield drop is observed (see Section 4). This
decrease in $d(\ln v)/d\tau$ is believed to be due to the operation of
a viscous drag at high rates, as discussed above.

At even higher rates of strain, the motion of dis
locations will be affecte by relativistic behaviour. This effect
is due to the distortion of the strain field of a dislocation as
its speed approaches that of elastic waves in the material. The
elastic wave velocity represents a limiting velocity at which
the drag on the dislocation becomes indefinitely large. A theoretical analysis of these effects has been given by Weertman [22].
The strain rates corresponding to such effects may be estimated
by substituting the elastic shear wave velocity c for v in (1.2.1)
together with a reasonable value, say $10^{11} m^{-2}$, for the mobile
dislocation density; this gives $\dot\gamma^p \simeq 10^5 s^{-1}$, i.e. a value one to
two orders of magnitude higher than that at which viscous drag

effects become significant. The stresses required to reach rates
of this order must therefore be very high, and are probably only
reached under explosive loading.

Chapter 2.

Wave Propagation.

2.1. Characteristics.

A review of the method of characteristics applied to the theory of stress waves in solids has been given by Hopkins [23]. The governing equations considered are of the form

$$(2.1.1) \qquad a_{ij} \frac{\partial u_j}{\partial x} + b_{ij} \frac{\partial u_j}{\partial t} + c_i = 0 ,$$

where the coefficients a_{ij}, b_{ij} and c_i are in general functions of the independent variables (x, t) and of the n dependent variables u_j; the repeated (dummy) suffix j implies summation over the values $j = 1, 2, ..., n$. It is found that there are curves C in the (x, t) plane across which there may be discontinuities in the first derivatives of u_i. These curves are given by

$$(2.1.2) \qquad dx = \lambda_\nu dt,$$

where the values of λ_ν are the roots of

$$(2.1.3) \qquad | a_{ij} - \lambda b_{ij} | = 0 .$$

Assuming that there are n distinct roots of (2.1.3), there are n families of characteristic curves C, along each of which the increments du_i are related by an equation of the type

$$(2.1.4) \qquad \begin{cases} A_i du_i + B dx = 0 \qquad \text{or} \\ \\ A_i du_i + \lambda B dt = 0, \end{cases}$$

where A_i and B are known functions of x, t and u_i.

Equations (2.1.2) and (2.1.4) can be used, in finite difference form, to construct a network of characteristics in the (x,t) plane, along which the integration may be carried out. If the quantities λ_ν depend only on x and t (2.1.2) may be integrated to give characteristics which are independent of the solution u_i; in the general case, however, λ_ν is a function of u_i and hence the characteristic network must be determined simultaneously with the solution itself.

2.2. Rate-Independent Material.

2.2.1. Longitudinal waves in a rod.

Let σ be the nominal or engineering axial stress, v the particle velocity and ρ the initial density of a uniform rod whose axis coincides with the axis Ox. Let x denote the initial position of a given particle (Lagrangian coordinate). Then the equation of motion is

$$\frac{\partial \sigma}{\partial x} + \rho \frac{\partial v}{\partial t} = 0 \ . \qquad (2.2.1)$$

Continuity of displacement requires that

$$\frac{\partial v}{\partial x} - \frac{\partial \varepsilon}{\partial t} = 0 \ , \qquad (2.2.2)$$

where ε is the engineering strain.

If stress components other than σ are neglected and the material is assumed to be rate-independent, the constitutive relation $f(\sigma_{ij}) = k$ reduces to

(2.2.3) $$\sigma = \Phi(\varepsilon) ,$$

so that

(2.2.4) $$\frac{\partial \sigma}{\partial t} - \Phi'(\varepsilon)\frac{\partial \varepsilon}{\partial t} = 0 .$$

      Equations (2.2.1), (2.2.2) and (2.2.4) constitute a set of the type (2.1.1), the three dependent variables being $u_1 = \sigma$, $u_2 = \varepsilon$, $u_3 = v$. The roots of (2.1.3) are

(2.2.5) $$\lambda = 0, \pm c$$

where

(2.2.6) $$c^2 = \Phi'(\varepsilon)/\rho .$$

      The first of (2.2.5) gives the particle paths $x = $ const., along which (2.2.3) holds; the other two roots correspond to wave propagation at speed c, a function of the tangent modulus $\Phi'(\varepsilon)$. Along these characteristics, (2.1.4) gives

(2.2.7) $$dv = \pm c\, d\varepsilon$$

and hence

(2.2.8) $$v = \pm \int c\, d\varepsilon .$$

From (2.2.6) and (2.2.7) we have

$$\sigma = \rho \int\limits_0^{\varepsilon} c^2 \, d\varepsilon = \rho \int\limits_0^{\upsilon} c \, d\upsilon, \qquad (2.2.9)$$

and hence the dynamic stress–strain relation can be determined experimentally if the wave velocity is measured as a function of strain or particle velocity in a unidirectional wave. This method has been used by Bell [24] Malvern [25] and others (Figs. 18, 19,20, see below and p.32). However, it has been shown in theoretical studies by Ripperger and Watson [26] that measurement of strain propagation velocities in initially unstressed specimens is not reliable as an indicator of the type of constitutive equation governing the material behaviour, since similar measurements would be obtained for several different assumed forms of equation.

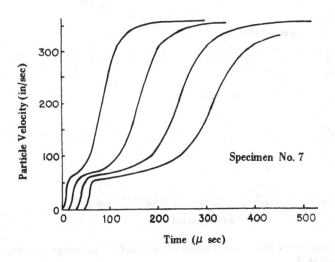

Fig. 18. Typical record of particle velocity vs. time [62]

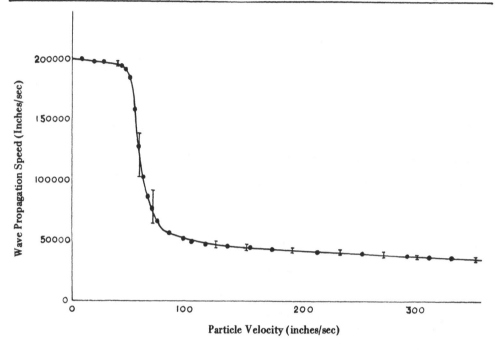

Fig. 19. Wave-propagation speed vs. particle velocity (based on averaged data from six velocity records) [62]

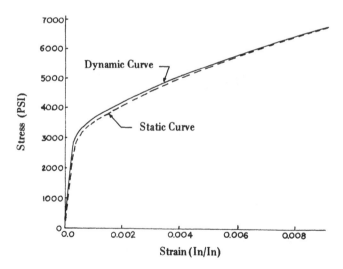

Fig. 20. Static and dynamic stress-strain curves (dynamic curve based on rate-independent theory and velocity tests) [62]

2.2.2. Radial shear waves in a cylinder or disc.

Let τ be the shear stress, γ the shear strain and v the tangential particle velocity for an element initially at radius r. Then assuming plane motion with cylindrical symmetry the equation of motion is

$$\frac{\partial \tau}{\partial r} - \rho \frac{\partial v}{\partial t} + \frac{2\tau}{r} = 0 \qquad (2.2.10)$$

while that of continuity is

$$\frac{\partial v}{\partial r} - \frac{\partial \gamma}{\partial t} - \frac{v}{r} = 0 . \qquad (2.2.11)$$

A rate-independent stress–strain relation $\tau = \Phi(\gamma)$ gives

$$\frac{\partial \tau}{\partial t} - \Phi'(\gamma) \frac{\partial \gamma}{\partial t} = 0 . \qquad (2.2.12)$$

Comparing $(2.2.10), (2.2.11), (2.2.12)$ with $(2.1.1)$ we see that in this case the quantities c_i are not all zero. The characteristics are given by $(2.2.5)$, where $c^2 = \Phi(\gamma)/\rho$. Along the curves $\lambda = \pm c$ $(2.1.4)$ gives

$$d\tau + 2\tau dr/r = \pm \rho c(dv - vdr/r) . \qquad (2.2.13)$$

Equations $(2.2.13)$ cannot be integrated except by a numerical procedure. Such an analysis for a linearly work-hardening material has been carried out by Rakhmatulin [27].

2.3. Rate-Dependent Material.

2.3.1. Longitudinal waves in a rod.

Equations (2.2.1) and (2.2.2) still apply, but (2.2.3) is replaced by a constitutive relation such as that of equation (1.1.9), which may be written

$$(2.3.1) \qquad \frac{\partial \sigma}{\partial t} - E \frac{\partial \varepsilon}{\partial t} + g(\sigma, \varepsilon) = 0$$

The analysis of the system (2.2.1), (2.2.2) and (2.3.1) was given by Malvern [28] The characteristic roots are again given by (2.2.5), where the propagation speed is now

$$(2.3.2) \qquad c = \pm (E/\rho)^{1/2},$$

i.e. the elastic wave speed.

Along the characteristics $\lambda = \pm c$, from (2.1.4)

$$(2.3.3) \qquad d\sigma = \pm \rho c \, dv - g \, dt$$

For a given function $g(\sigma, \varepsilon)$, (2.3.3) may be integrated numerically along the fixed characteristic curves $dx = \pm (E/\rho)^{1/2} dt$. Examples of such computations have been given by Cristescu [29] for the case where

$$g(\sigma, \varepsilon) \quad \propto \quad \sigma - \Phi(\varepsilon),$$

$\Phi(\varepsilon)$ being the static stress-strain function. Dynamic stress-strain relations were calculated for various sections of a rod subjected at one end to a velocity increasing and decreasing linearly with time (Figs. 21,22, see below and p.36). It was assumed in the computation that (2.3.3) applies for $g(\sigma,\varepsilon)>0$, i.e. for $\sigma>\Phi(\varepsilon)$. This means that plastic flow continues for some time after the stress starts to decrease.

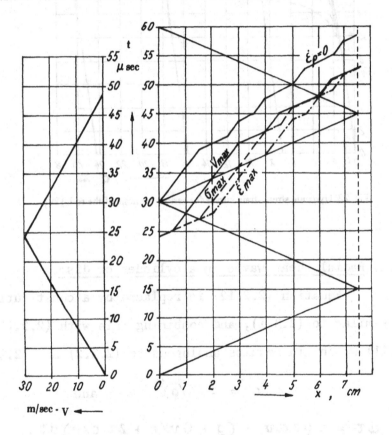

Fig. 21. Characteristic field in which the following lines are represented: loading/unloading boundary $(\dot{\varepsilon}_p = 0)$, maximum stress (σ_{max}), maximum particle velocity (υ_{max}), and maximum rate of strain $(\dot{\varepsilon}_{max})$. [29]

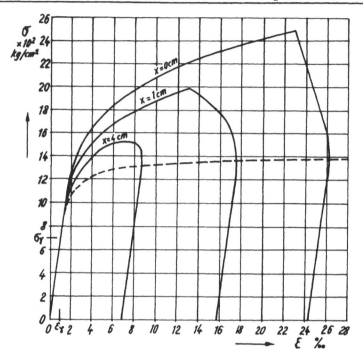

Fig. 22. Dynamic stress-strain curves for various sections of the rod. [29]

2.3.2. Radial shear waves in a cylinder or disc.

Equation (2.2.12) is replaced by a constitutive
relation similar to (2.3.1), and combining this with (2.2.10)
and (2.2.11) we obtain results analogous to (2.3.2) and (2.3.3):

$$(2.3.4) \qquad\qquad c = \pm (G/\rho)^{1/2} \qquad\qquad \text{and}$$

$$(2.3.5) \quad d\tau = \pm \rho c\, dv - (g + Gv/r + 2\tau c/r)\, dt .$$

Here c is the elastic shear wave speed, G is the shear modulus
and $g(\tau,\gamma)/G$ is the plastic shear strain rate.

2.4. Incremental Waves.

Consider a body which is in a plastic state caused by a preexisting load, either static or dynamic. If a small incremental load is applied, waves will be propagated and the wave motion will depend on two factors: Firstly, whether the incremental load is of the same sign as that of the initial load; secondly, whether the material is rate-dependent or rate-independent in its mechanical behaviour. If the incremental load is of opposite sign to the initial load, the disturbance will be propagated at the elastic wave speed, assuming that the material unloads elastically; if it is of the same sign, however, the speed of propagation in a rate-independent material will be less than the elastic wave speed. This may be shown by a perturbation method [23]; for uniaxial tension or compression loading, the wave speed is

$$ \sigma = [\Phi'(\varepsilon_{o})/\varrho]^{1/2}, \qquad (2.4.1) $$

where ε_{o} is the pre-strain. For values of ε_{o} appreciably greater than the yield strain, therefore, the disturbance will be propagated at a speed very considerably less than the elastic wave speed. Experimental tests involving such incremental loading have been carried out by a number of workers [30 – 34] Many of these tests have indicated that the propagation speed is that of elastic waves, in contrast to that predicted by (2.4.1).

Such a result is expected if the material behaviour is governed by an equation such as (2.3.1). In this case, a sharp-fronted wave travels with speed $(E/\rho)^{1/2}$; the amplitude does not remain constant, however, but decays due to the plastic flow of the material.

The use of the one-dimensional theory in this context has been criticized by Craggs [35]. An approximate analysis by De Vault [36] shows that the neglect of lateral motion in the rod leads to an underestimate of the wave speed. To avoid this difficulty, tests have been made on thin-walled tubes under torsional loading [37 - 39]. These tests all showed that the wave front was propagated at a speed much greater than that given by the rate-independent theory based on the static stress-strain curve; in many cases, the velocity of the front was equal to that of elastic waves, within experimental accuracy (Figs. 23 - 27, see pp. 39 - 42).

Experiments have also been made on short tubular specimens to determine directly the dynamic stress-strain curve under incremental loading [39]. These indicate that the initial response to such loading is essentially elastic (see Section 3.1).

2.5. Shock Waves.

Under certain conditions stresses may be propagated in the form of shock waves, across which the stress, strain and particle velocity are discontinuous. Such waves can occur

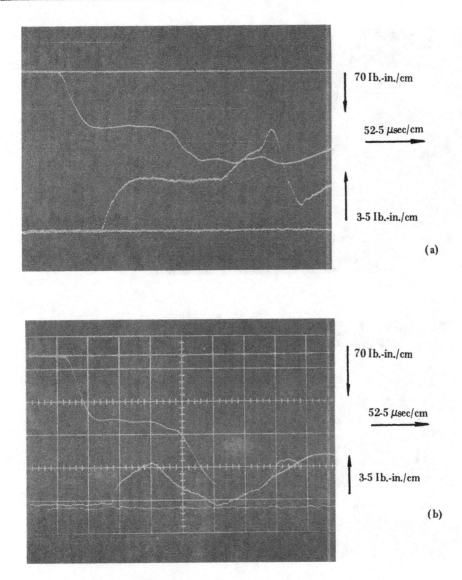

Fig. 23. Oscillograms for wave tests on copper.
(a) zero pre-strain ; and
(b) pre-strain 10·0 per cent.

when the wave speed c increases with increasing stress or strain,

so that the characteristic curves converge.

The front of a propagating wave then becomes increasingly steep

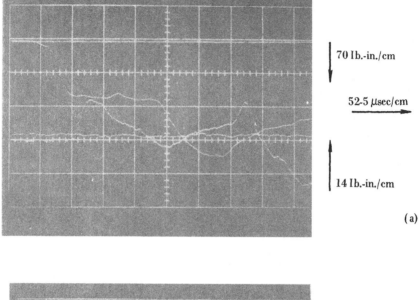

70 lb.-in./cm

52-5 μsec/cm

14 lb.-in./cm

(a)

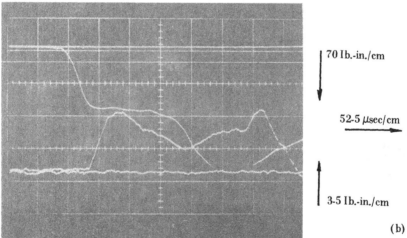

70 lb.-in./cm

52-5 μsec/cm

3-5 lb.-in./cm

(b)

Fig. 24. Oscillograms for wave tests on aluminium.
(a) pre-strain 1-9 per cent ; and
(b) pre-strain 16-8 per cent.

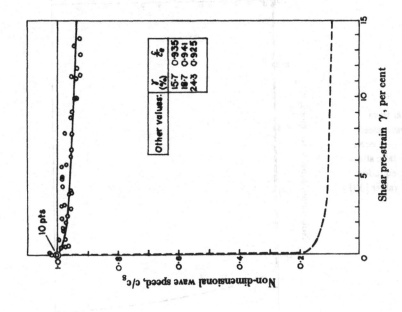

Fig. 26. Ratio of incremental wave speed c to elastic wave speed c_e, for copper. —— curve derived from quasi-static stress-strain relation using equation (1); ○ experimental values [39]

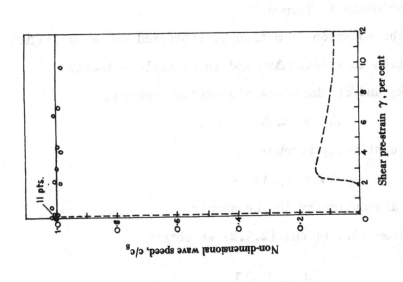

Fig. 25. Ratio of incremental wave speed c to elastic wave speed c_e, for mild steel. —— curve derived from quasi-static stress-strain relation using equation (1); ○ experimental values. [39]

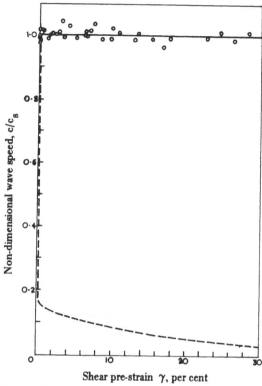

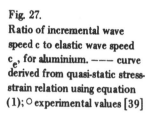

Fig. 27.
Ratio of incremental wave
speed c to elastic wave speed
c_e, for aluminium. —— curve
derived from quasi-static stress-
strain relation using equation
(1); ○ experimental values [39]

so that a discontinuity is formed.

The equation of motion is expressed in terms of (Δv)
the discontinuities in stress $(\Delta \sigma)$ and in particle velocity
across the shock, and of the shock propagation speed c_s :

(2.5.1) $\Delta \sigma + \rho c_s \Delta v = 0 .$

The equation of continuity becomes

(2.5.2) $\Delta v + c_s \Delta \varepsilon = 0 .$

where $\Delta \varepsilon$ is the discontinuity in the strain.

From (2.5.1) and (2.5.2) we obtain

(2.5.3) $c_s^2 = \dfrac{1}{\rho} \dfrac{\Delta \sigma}{\Delta \varepsilon} .$

Equations (2.5.1), (2.5.2) and (2.5.3) may be

applied to the propagation of a wave under conditions of uniaxial stress or of uniaxial strain. Much of the experimental work
on shock waves on solids has been carried out in uniaxial strain
conditions with very high compressive stresses produced by explosives. Under these conditions the stress-strain relation is ef-
fectively linear up to the dynamic compressive yield stress p_y
but for higher stresses the relation is such that $d\sigma/d\varepsilon$ increases
with increasing compressive strain. Thus for applied pressures p_A
somewhat greater than p_y , two shock fronts are propagated, a precursor of amplitude p_y travelling at the appropriate elastic wave
speed, and a slower plastic wave of amplitude p_A . For still higher values of p_A the plastic wave speed may exceed the elastic
wave speed, so that a single shock of amplitude p_A is propagated.

        The dynamic elastic limit p_y may be determined by
measuring the particle velocity associated with the precursor
wave. This is most conveniently done by allowing the wave to re-
flect from a free surface and determining the velocity imparted
to the surface. Values obtained in this way for metals may be
two or three times those found under static loading [40, 41] . It
has also been found that the amplitude of the precursor p_y decreases as it propagates (Fig. 28). This effect has been explained in terms of a rate-dependent constitutive relation based on
dislocation dynamics [42] .

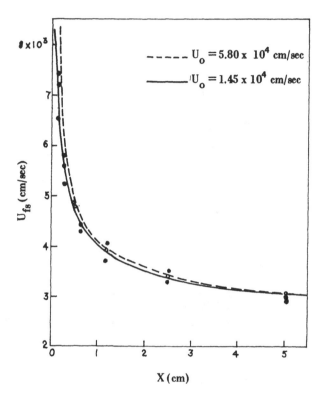

Fig. 28. Cartesian plot of the Hugoniot elastic wave free-surface velocity for annealed Armco iron. The parameter u_o is the initial impact velocity. [42]

Chapter 3.
Experimental Methods and Results.

3.1. Tests at Medium Strain Rates.

The problem involved in measuring the mechanical properties of rate-dependent materials have been discussed by Cooper and Campbell [43]. The observed behaviour depends on the interaction of the specimen and test machine, which is largely determined by their relative stiffnesses. If the stiffness of the test machine is much less than that of the specimen ('soft' machine), the load-time history may be controlled; if it is much greater ('hard' machine), the strain-time history may be controlled.

Clark and Wood [44] were the first to employ a soft machine to study the time dependence of yielding in steel. They found that if a fixed load was applied to a tension speci-men, in a time of the order of 10 ms, a finite 'delay period' elapsed before gross yielding was observed. This delay period was found to be a function of the applied stress, decreasing rapidly as the stress was increased above the static yield stress (Fig. 29, see p.46). The influence of temperature was also investigated, and it was shown by Cottrell [45] (Fig. 30, see p.47) that the results were reasonably consistent with a thermal-activation rate equation of the form

(3.1.1) $t_d = t_o \exp[U(\frac{\sigma}{\sigma_o})/kT]$.

In (3.1.1), t_d is the delay period for an applied stress σ, t_o and σ_o are constants, k is Boltzmann's constant and T is the absolute temperature; $U(\sigma/\sigma_o)$ is the activation energy for the mechanism controlling the movement of dislocations. Originally, this mechanism was believed to be the release of dislocations from pinning atoms of carbon or nitrogen [45, 46, 47]. However, later work [48, 49] indicated that the mechanism was one of dislocation motion rather than release.

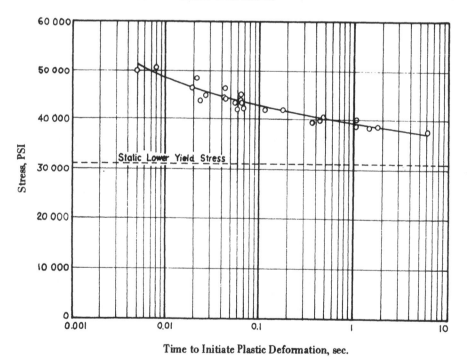

Fig. 29. Stress *versus* Time to Initiate Plastic Deformation for 0.19 per cent Carbon Annealed Steal. [44]

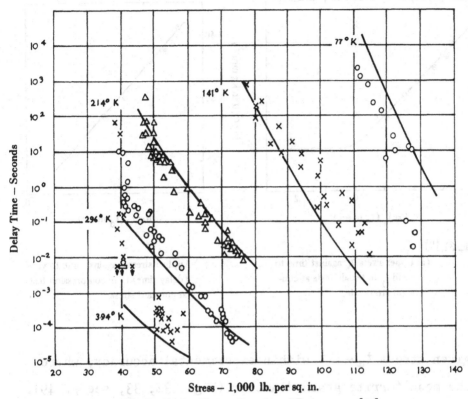

Fig. 30. *Delay times for Yield in Steel at Various Stresses and Temperatures* [45]
The experimental points were obtained by Clark and Wood (1949). The theoretical curves are
derived from equation (3.1)

If the applied stress increases with time, yield

occurs at a stress σ_y after a time t_y, and it has been shown [9,50]

that the results of various types of test may be correlated by

assuming that gross yielding (Lüders band formation) occurs when

the gradients of dislocation density reach a critical value (Fig.

31, see p.48).

The effect of grain size on the delay period for

mild steel has also been determined [51]   and it is found that

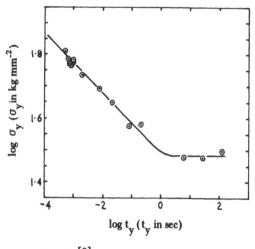 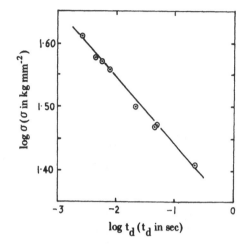

Fig. 31. [9]

(a) Upper stress σ_y against time to
yield t_y in tensile tests at constant stress rate.

(b) Applied stress σ against time to yield
(delay time) t_d in compression tests
at constant stress.

at a given stress $t_d \propto d^{-3}$ within experimental accuracy, where
d is the mean ferrite grain diameter (Figs. 32, 33, see p. 49).
From the straight lines of Figs. 31, 32 and 33, together with
the thermal—activation rate equation (3.1.1), in which the function U is approximated by the equation [52]

(3.1.2) $$U = U_0 \ln(\sigma_0/\sigma) \, ,$$

we obtain

(3.1.3) $$\sigma = \sigma_0 (A/d^3 t_d)^{kT/U_0},$$

where A and U_0 are constants. At room temperature $kT/U_0 \simeq 1/9$ so
that (3.1.2) becomes

(3.1.4) $$\sigma \, d^{1/3} t_d^{1/9} = \text{const.}$$

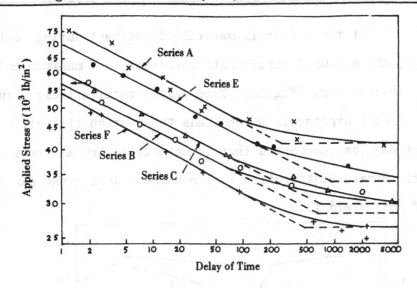

Fig. 32. Applied stress against delay time for grain sizes A, B, C, E, F. [51]

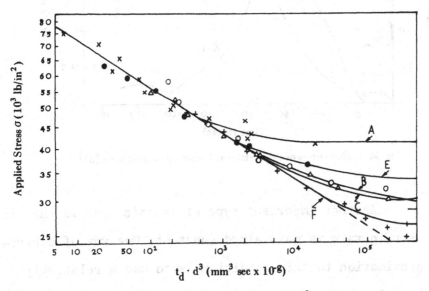

Fig. 33. Applied stress against delay time X (mean grain diameter)^3 for grain sizes A, B, C, E, F. [51]

If the stress is maintained constant during post-yield flow, the measured strain rate increases to a maximum and decreases towards zero (Fig.34). This agrees qualitatively with the 'overstress' hypothesis which leads to equation (1.1.8); however it must be remembered that in materials such as steel the plastic strain distribution is non-uniform during the early stages of post-yield flow.

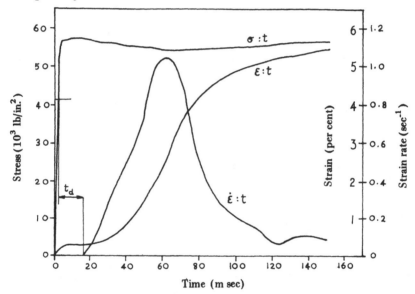

Fig. 34. Typical curves of stress, strain and strain rate against time. [51]

Another important type of dynamic test is that in which the strain rate is maintained constant. One way of achieving an approximation to this condition is to use a relatively stiff machine in which the imposed speed of deformation is constant during the test. In any real machine, a finite time is required to accelerate the moving parts to the desired speed, but

by the use of a quick-acting release valve, it is possible to re
duce this time to about 2 ms [43] (Fig.35).

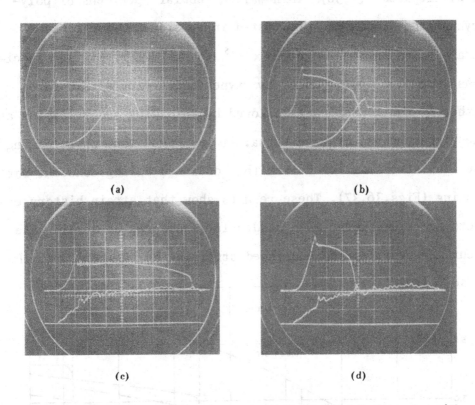

(a)   (b)

(c)   (d)

Fig. 35. Tensile test oscillograms obtained at medium strain rates. (a) Stress-time (both traces) ; total time
base for upper trace 20 msec, for lower trace 5 msec ; temperature, 200° K. (b) Trace data as for (a) ; temperature, 163° K. (c) Stress-time (upper trace), strain rate-time (lower trace); total time base for both traces
10 msec.; temperature, 250° K. (d) Trace data as for (c); temperature, 100° K. Stress trace shows evidence
of twinning before yield.

Some experimental work has been done at medium
strain rates under stress systems other than simple tension or
compression. This includes investigations of flexure [53], combined tension and shear [4,54], and pure shear [55] . The latter
work indicates that iron is more strain-rate sensitive in shear

than in tension or compression.

The effect of strain history has been investigated by Klepaczko [7,56]. Thin-walled tubular specimens of polycrystalline aluminium were tested in torsion at five constant strain rates in the range $1 \cdot 6 \times 10^{-5}$ to 0.624 s^{-1}. Similar specimens were then tested under two types of varying strain rate: either at the lowest rate followed by unloading and reloading at the highest rate, or vice versa. It was found that on reloading the flow stress differed from that obtained during constant rate testing (Figs.36,37). These results show that strain history effects are significant when sudden large changes of strain rate occur. However, during continued straining at the new rate, the

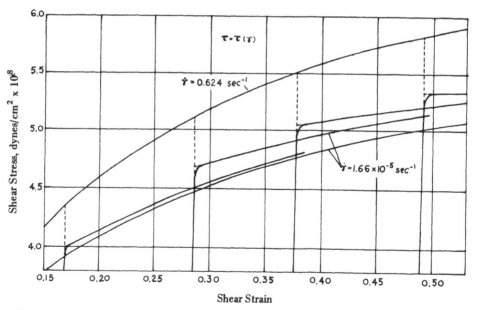

Fig. 36. The results of strain rate change for four initial strain strains; initial strain rate is $\dot{\gamma}_i = 0.624$ sec^{-1}, and strain rate of reloading is $\dot{\gamma}_r = 1.66 \times 10^{-5}$ sec^{-1}. [56]

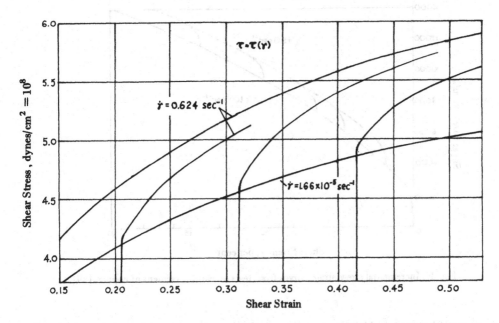

Fig. 37. The results of strain rate change for three initial values of strain; initial strain rate is $\dot{\gamma}_i = 1.66 \times 10^{-5}$ sec^{-1}, and strain rate of reloading is $\dot{\gamma}_r = 0.624$ sec^{-1}. [56]

flow stress tends towards that corresponding to a constant-speed test at this rate; that is, the 'memory' of the earlier deformation fades with continued straining. The physical explanation for the history effect is believed to lie in a variation in the dislocation structure with strain rate, a smaller average network size being found after straining at high rate. Similar effects were noted by Campbell and Dowling [39] in tests during which the shear strain rate was suddenly increased from 5×10^{-3} to 90 s^{-1} (Fig.38, see p.54).

Recent work on the B.C.C. metals niobium and molybdenum [57] has shown that history effects may be quite large,

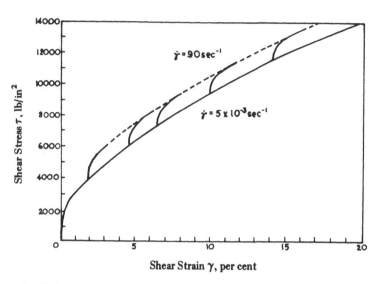

Fig. 38. Incremental stress-strain curves from tests on short specimens of copper. [39]

and significantly different from those observed in aluminium. In both niobium and molybdenum, it is found that a sudden increase in the strain rate may be accompanied by an increase in the flow stress to a value considerably higher than that corresponding to a constant-speed test at the higher rate (Figs.39,40, see p.55). The behaviour is complex, however, since for certain values of initial and final strain rate the response is qualitatively similar to that found by Klepaczko for aluminium (Figs.41,42, see p.56).

The effects of strain history under biaxial loading have been studied by Lindholm [4], in experiments on thin-walled tubular specimens subjected to combined axial and twisting deformation. In these experiments, it was found that the orientations of the stress and strain-rate tensors were the same

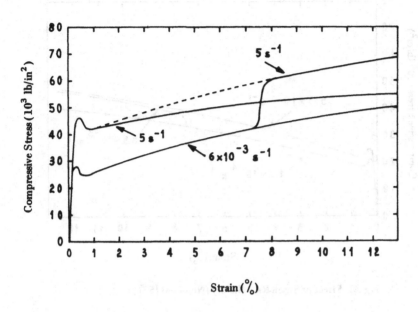

Fig. 39. Effect of Strain-Rate History (Niobium) [57]

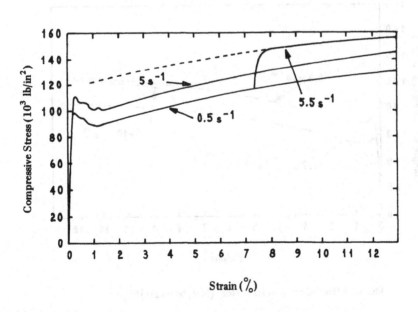

Fig. 40. Effect of Strain-Rate History (Molybdenum) [57]

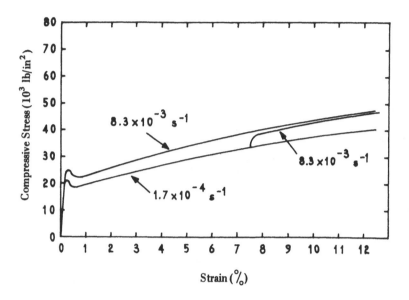

Fig. 41. Effect of Strain-Rate History (Niobium) [57]

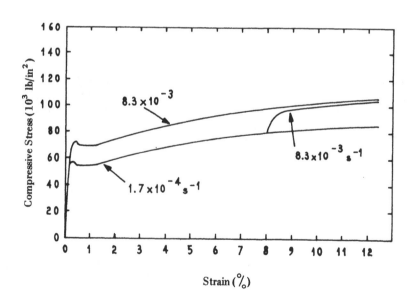

Fig. 42. Effect of Strain-Rate History (Molybdenum) [57]

within the accuracy of measurement (Figs.43,44, see below and
p.58). Assuming that the elastic strain rate was negligible com-
pared with the plastic strain rate, this accords with the constitutive relation (1.1.7). In a test in which both the rate and
direction of loading were suddenly and simultaneously changed,
it was found that the flow stress after the change was the same

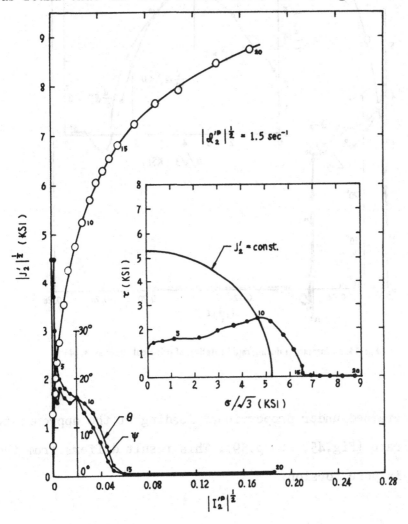

Fig. 43. Deformation during gradual rotation of principal stress axes. [4]

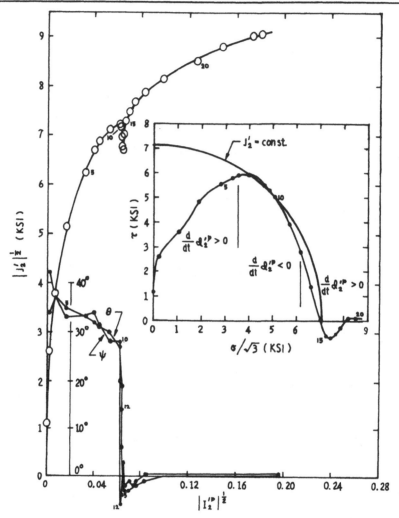

Fig. 44. Deformation during rapid rotation of principal stress axes. [4]

as that obtained under proportional loading at the appropriate
constant rate (Fig.45, see p.59). This result differs from those
obtained in pure torsion.

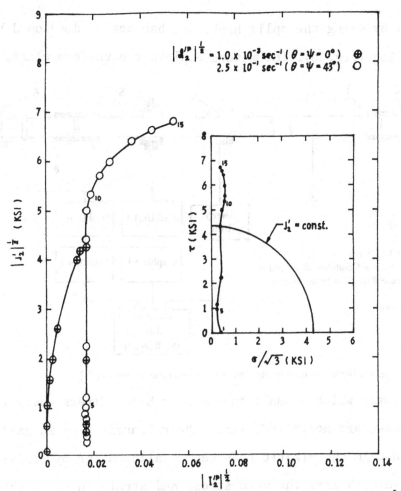

Fig. 45. Deformation during change in rate and direction of loading after unloading. [4]

3.2. Tests at High Strain Rates.

When the time of application of the load is reduced to a value comparable to the time required for an elastic
wave to traverse the parts of the apparatus through which the
load is applied, wave propagation in the apparatus becomes important. t is then necessary to design the apparatus in such a
way that the wave motion in it can be analysed. This can be

achieved by using the split Hopkinson–bar method developed by

Kolsky [58] (Fig.46). The method employs two uniform elastic

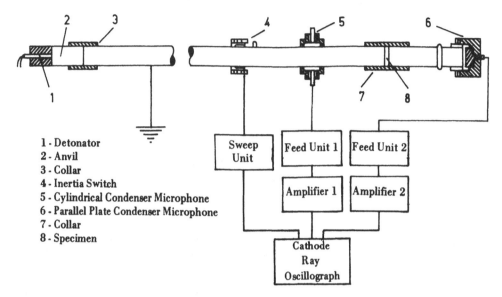

1 - Detonator
2 - Anvil
3 - Collar
4 - Inertia Switch
5 - Cylindrical Condenser Microphone
6 - Parallel Plate Condenser Microphone
7 - Collar
8 - Specimen

Fig. 46. Davies bar used for dynamic stress-strain measurements . [58]

rods, between which a small specimen is held. Stress waves in

the specimen are neglected, since their transit time is small,

while those in the elastic rods can be allowed for by applying

elastic wave theory. The mean stress and strain in the specimen

can thus be obtained as functions of time from measurements made

of the waves propagated in the two rods.

This method of test has been used for compression

loading [58 – 66] and has also been adapted for tension [67 –

– 69]. With both these types of loading, difficulties arise be-

cause of end effects and lateral strains in the specimen; with

tension loading, plastic instability (necking) may occur at rel-

atively small strains (Fig.47, see p.61). All these effects are

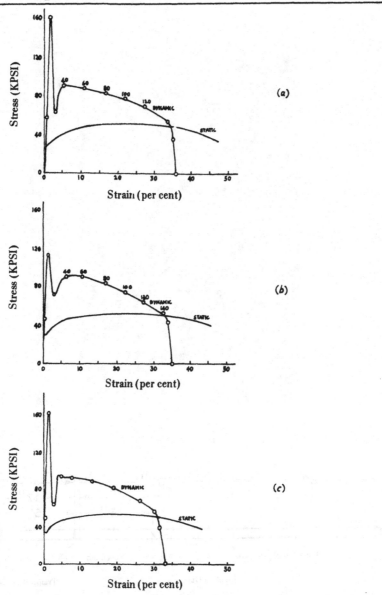

Fig. 47.
Stress-strain curves of steel specimens (batch 3): (a) uniradiated, (b) radiation dose 8 x 10[17] neutrons/sq cm, (c) radiation dose 5 x 10[18] neutrons/sq cm. Times microseconds are shown on dynamic curves [63]

eliminated or greatly reduced by employing torsional loading of

a thin-walled tubular specimen, and the method has recently been

adapted for such loading [39,70,71] (Figs.48 - 52, see pp.62 - 66).

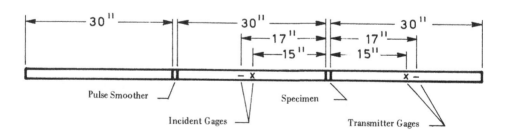

Fig. 48. Hopkinson tube for torsion test. The photograph shows details of the loading end with the sheet explosive and the detonator in place. The schematic diagram shows the location of the gages.

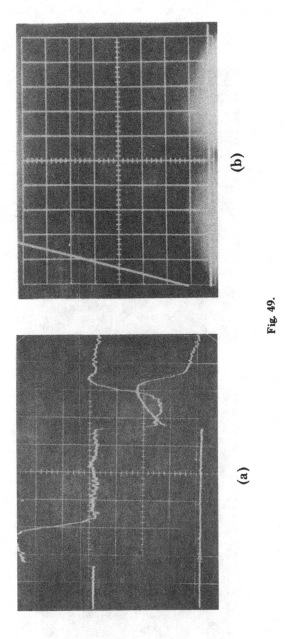

Fig. 49.

(a)

(b)

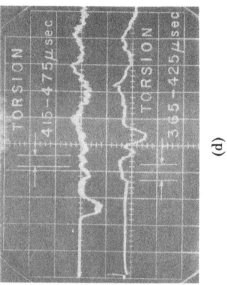

(c)

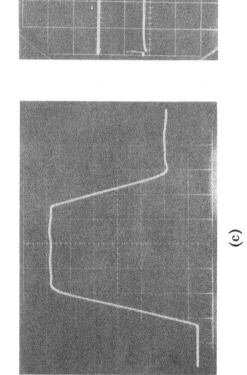

(d)

Fig. 49. Dynamic torsion test No. 3. Oscilloscope records : (a) torsion gages on Hopkinson tube, (b) calibration of integrator, (c) integrated output of torsion gages on incident tube, (d) axial gages on incident tube, top trace is compession, bottom is bending. Sweep speed 50 µsec/ div. in (a), (b), (c), 100 µsec/div.

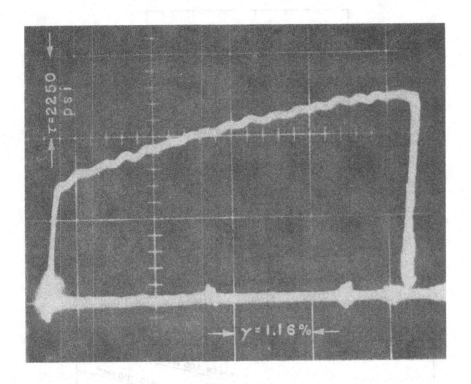

Fig. 50. Enlargement of oscilloscope record showing stress versus strain for dynamic torsion (test No 3).

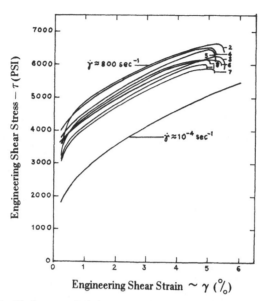

Fig. 51. Stress-strain behaviour in shear for 1100—0 alluminium alloy.

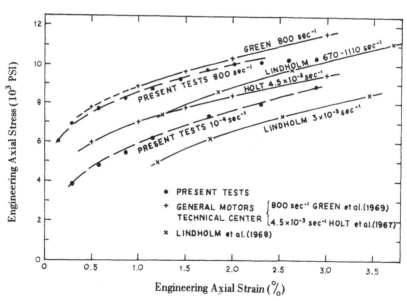

Fig. 52. Comparison with results obtained by other investigators in axial tests on 1100—0 alluminium alloys. The shear stress and strain from the present tests are converted to axial stress and strain by using the relations $\sigma = \sqrt{3}\tau$ and $\epsilon = \gamma/\sqrt{3}$.

The use of torsional pulses in the measuring bars has the further advantage that propagation is free of the geometric disper-
sion which affects longitudinal waves.

In general, it is
found that the
flow stress at a
given strain in-
creases with
strain rate,
though for high-
strength alumin-
ium alloys the
fractional in-
crease is very
small [66]. In
many cases the
logarithmic rate
sensitivity of
the flow stress
$[\partial \sigma / \partial \ln \dot{\varepsilon}]_{T,\varepsilon}$
is constant over
a large range of
strain rates, im
plying that the

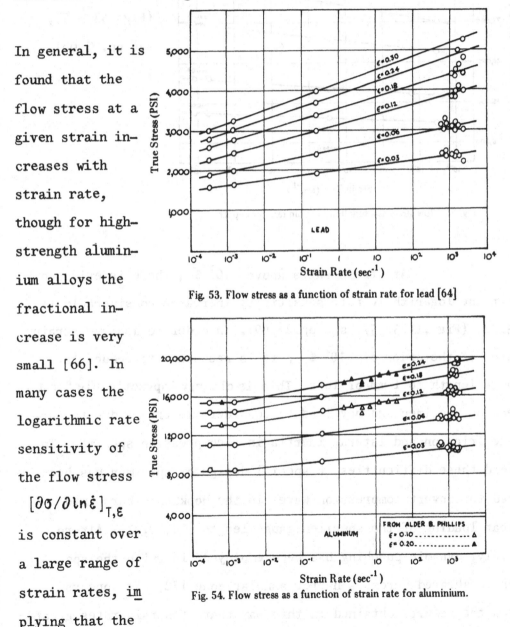

Fig. 53. Flow stress as a function of strain rate for lead [64]

Fig. 54. Flow stress as a function of strain rate for aluminium.

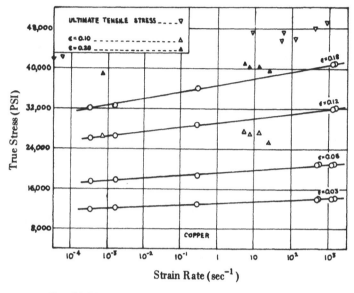

activation vol-
ume is indepen-
dent of stress
(Figs.53 – 55,
see pp.67,68).

Fig. 55. Flow stress as a function of strain rate for copper.

At strain rates above 10^3 s^{-1}, there is evidence
that the logarithmic rate sensitivity increases considerably [15,
65,72] (Figs.13,56,57, see pp.22,69). In order to achieve strain
rates of the order of 10^4 s^{-1}, it is necessary to reduce the
gauge length to 1 mm or less. This is clearly impracticable for a
tension specimen and may lead to considerable errors due to surface friction and lateral inertia in a compression specimen. To
avoid these difficulties, a special design of specimen has been
used to convert compression waves in the Hopkinson bars into
shear loading of the specimen gauge length [15,20,21]. Alterna-
tively, dynamic punching has been employed, in which the specimen is sheared across a narrow annular zone [72]. It appears
from the results obtained in this way that at strain rates great

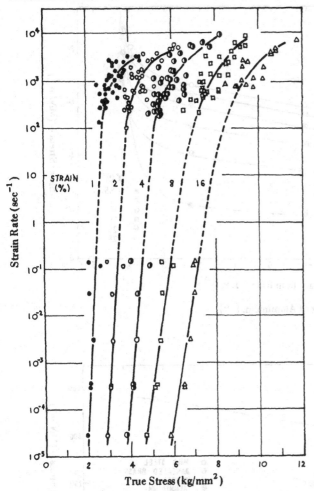

Fig. 56. Relation between the logarithm of the strain rate and the stress for polycrystalline aluminium, including the data for quasistatic deformation. [65]

er than about $10^3 s^{-1}$, the flow stress tends to increase linearly with strain rate, rather than with its logarithm (Figs. 14,15, 16,58, see pp. 23,24 and below). This behaviour has been attributed to the action of a viscous drag force opposing the motion of dislocations (see section 1.2). The following table gives measured values of the macroscopic viscosity coefficient η defined by equation (1.2.6), in units of $kNsm^{-2}(10^4$ poise) [72]:

	Aluminium	Copper	Brass	Mild Steel	Zinc
Poly-crystalline	2.1	3.6	5.5	2.8, 2.1	–
Single crystals	1.2	10.8	–	–	0.5

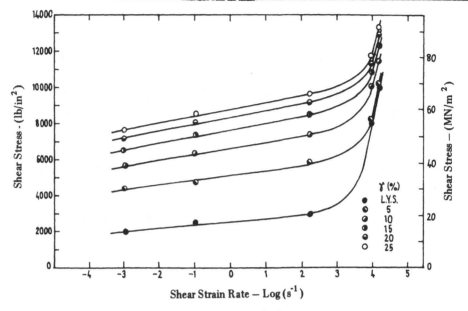

Fig. 57. Strain-Rate Sensitivity of Aluminium. [72]

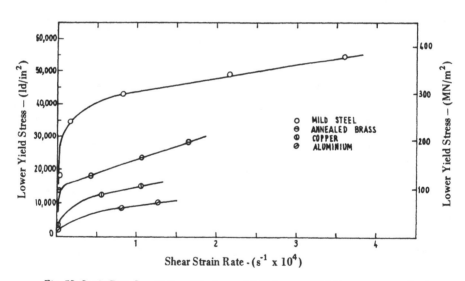

Fig. 58. Strain-Rate Sensitivity of the Lower Yield Stress at High Rates of Strain. [72]

These values may be ralated to measurements which
have been made of the dislocation damping constant B. This constant has been determined for copper and aluminium by measuring
the velocity of individual dislocations as a function of the applied stress [73,74]. If ϱ_m is the density of mobile dislocations
and v is their mean velocity, the total energy dissipated in unit
volume of material is $B\varrho_m v^2$ and this may be equated to the rate
at which the viscous component of stress $\eta \dot{\gamma}^P$ does work:

$$B\varrho_m v^2 = \eta(\dot{\gamma}^P)^2. \qquad (3.2.1)$$

Combining (3.2.1) with the kinematic relation
(1.2.1) we obtain

$$\varrho_m = B/\eta b^2. \qquad (3.2.2)$$

From the measured values of η and B we obtain $\varrho_m \simeq 3 \times 10^{11} \, m^{-2}$
for copper and $1.5 \times 10^{11} m^{-2}$ for aluminium at the lower yield
strain. As noted in section 1.2, such values for the mobile dislocation density imply that the dislocation velocity will not
approach the elastic wave speed until strain rates of the order
$10^5 s^{-1}$ are reached.

An interesting observation which appears to be re-
lated to the increase in rate sensitivity at rates above $10^3 S^{-1}$
is that a drop of stress at yield may occur in materials which
do not show a drop at lower rates of strain. Such yield points
have been found in copper-antimony [75], pure copper [76] and

other metals [72] . Combining (1.2.9) with (1.2.1), the condition
for local plastic instability becomes

(3.2.3) $$\frac{d}{d\gamma^p}\left[\ln\left(\rho_m\,\frac{d\rho_m}{d\gamma^p}\right)\right] > G\,\frac{d}{d\tau}\left(\ln\dot{\gamma}^p\right)\ .$$

At high rates, the right side of (3.2.3) decreas-
es so that the inequality is satisfied; the plastic strain then
becomes localized and a yield drop occurs.

Chapter 4.
Applications.

4.1. Metal Forming.

Many metal-forming processes involve plastic deformation at medium or high rates of strain. In most such processes the material is subjected to a complex stress and strain history as it is deformed, and it is therefore difficult to determine precisely the strain rates involved. However, estimates of the orders of magnitude of these rates may be obtained by making reasonable assumptions concerning the kinematics of deformation. Atkins [77] has derived the following values for typical dimensions and speeds of working: sheet, rod or wire drawing, $1-10^3 s^{-1}$; cold rolling, $10^2-10^3 s^{-1}$; deep drawing, $1 s^{-1}$. Green and Maiden [78] have estimated that strain rates in deep drawing may be as high as $10^2 s^{-1}$.

In metal-forming processes, the plastic strains reached are very large, so that data obtained in simple tension or compression are inadequate because of the limited strains that can be achieved. One method of circumventing this difficulty is to test, at medium or high rate, specimens which have previously been cold-worked to a given strain. This method has been used by Karnes and Ripperger [79] for aluminium, and by Atkins and Porter [80] for mild steel (Figs.59,60, see p.74). Such an approach is of course only strictly valid if strain history effects are negligible.

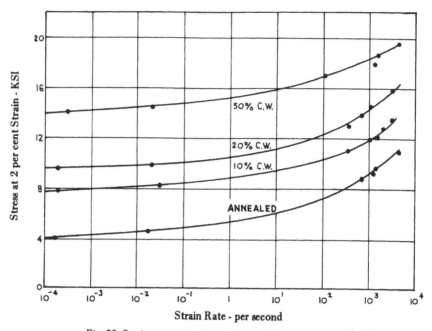

Fig. 59. Strain rate sensitivity of stress at 2 per cent strain. [79]

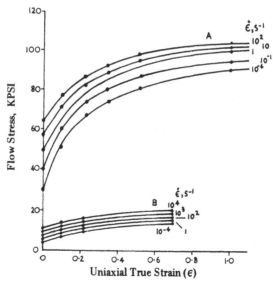

Fig. 60. Stress/strain curves replotted as flow curves at constant rate. [77]
    A *mild steel; data from Atkins and Porter.*
    B *aluminium; data from Karnes and Ripperger.*

The strain rates involved in machining have been estimated by Stevenson and Oxley [81] . By the measurement of grids inscribed on the work-piece, they were able to deduce the velocity gradients, i.e. strain rates, in the deformation zone; it was found that the mean shear strain rate in the zone was given by

$$\dot{\gamma}_m = 1.10 \, V_s/d ,\qquad\qquad (4.1.1)$$

where V_s is the shear velocity and d is the depth of cut. For $V_s = 5 \, ms^{-1}$ and $d = 0.05 \, mm$ therefore, the value of $\dot{\gamma}_m$ is of order $10^5 \, s^{-1}$. By measurement of the tool force it is possible to estimate the flow stress as a function of strain rate $[82,83]$. results obtained in this way agree well with extrapolation from values obtained at lower rates (Fig.61, see p.76). The rapid increase in the logarithmic rate sensitivity at high rates corresponds to that observed in shear tests using the split Hopkinson-bar technique.

Punching or blanking may also give rise to very high strain rates [84 – 88] . By measurement of the punch force and displacement it is possible to determine stress-strain curves at rates from 10^{-4} to $10^4 \, s^{-1}$ (Figs.62,63, see p.77). It is found that although the flow stress increases as the strain rate is increased, the energy absorbed in punching remains approximately constant; this is because the displacement to rupture decreases with increasing speed of deformation.

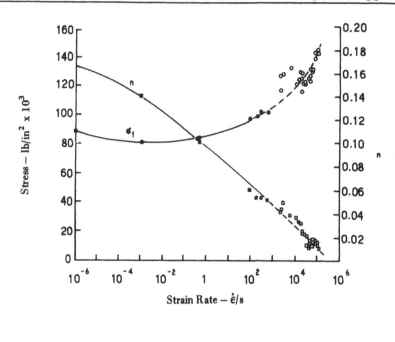

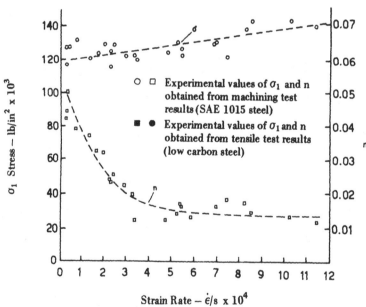

Fig. 61. Flow stress characteristics [83]

$$(\sigma = \sigma_1 \epsilon^n)$$

○ □   Experimental values of σ_1 and n obtained from machining test results (SAE 1015 steel)

● ■   Experimental values of σ_1 and n obtained from tensile test results (low carbon steel)

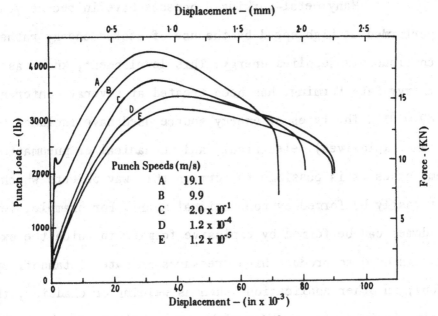

Fig. 62. Load-Displacement Curves for Copper [72]

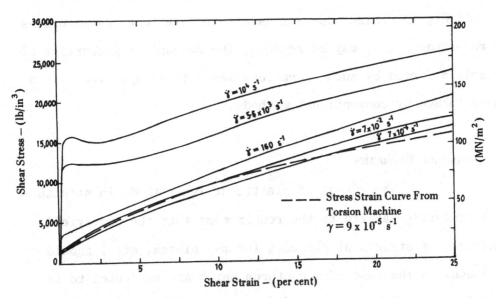

Fig. 63. Shear Stress-Strain Curves for Copper [72]

Many metal-working processes have in recent years been performed at high speed by the use of stored energy rather than continuously supplied energy. This development, known as High Energy Rate Forming, has been treated at several conferences [89 – 91]. The types of energy source used have included chemical (explosives), electrical, and mechanical or pneumatic. In some cases it is possible to form in this way objects which cannot easily be formed by conventional means. For example, very large domes can be formed by explosive forming in which the explosive is used to produce high pressures in water (standoff operation); in other applications such as welding or cladding, the explosive is in direct contact with the workpiece. In standoff operations the strain rates produced are typically of the order of $10^2 \, s^{-1}$; in contact operations strain rates one or two orders of magnitude higher may be reached. The mechanical properties of materials formed by such operations may differ appreciably from those formed by conventional methods.

4.2. Structural Mechanics.

The advent of plastic design methods in structural engineering has led to the requirement that the post-yield behaviour of structural elements (beams, plates, etc.) should be known. In the case of structures which are subjected to impact loading, this means that the dynamic plastic behaviour must be determined. Cristescu [29] has considered the mechanics of

wires and membranes under dynamic loading. The strength of beams
in rapid plastic flexure has been measured by Rawlings [92] and
by Aspden and Campbell [53] . The response of beams, frames and
plates under impact loading has also been studied [93 - 102] .
These investigations have shown that in general it is necessary
to allow for the strain rate in determining the dynamic plastic
behaviour of structures. For mild-steel and aluminium-alloy

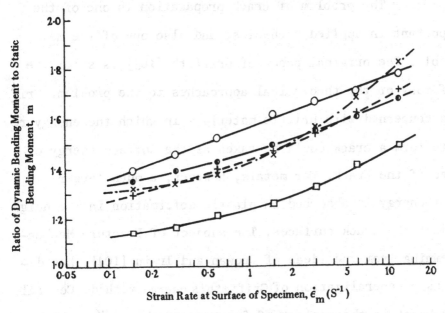

Fig. 64.   Variation of upper and lower yield moments with strain rate at surface of specimen.
O, Upper and o, lower yield moment derived from compression results; x, upper and □, lower
yield moment from flexure tests. + , Empirical curve, $m = 1 + 0.434\,\epsilon 1\beta_m$. [53]

beams in pure bending, an empirical relation due to Bodner and
Symonds [94] has been found to be adequate as a first approximation at least:

$$\dot{k} = C(M/M_0 - 1)^p .$$

$$(4.2.1)$$

In (4.2.1), κ is the curvature, M is the bending moment, M_0 is
the 'fully-plastic' or 'limit' moment, and C, p are constants
for a given material; the equation therefore neglects work-hardening. Fig.64 (see p.79) compares the predictions of (4.2.1)
with experimental results for mild steel.

4.3. Crack Propagation.

The problem of crack propagation is one of the
most important in applied mechanics, and also one of the most
intractable. The original paper of Griffith [103] is still the
basis of many of the theoretical approaches to the problem. Grif
fith was concerned with brittle materials in which the energy re
quired to form a crack could be taken as the surface energy of
the faces of the crack. For metals, however, a much larger
amount of energy is absorbed in plastic deformation in the neigh
bourhood of the crack surfaces. The subject of Fracture Mechanics, stemming from the ideas of Orowan and Irwin [104], has developed as a generalization of Griffith's energy method. Central
to the subject is the concept of fracture toughness K_c, which is
related to the energy absorbed by plastic deformation in extend-
ing the crack. The critical applied stress at which a crack of
length a will propagate is given by

$$(4.3.1) \qquad\qquad \sigma = \beta K_c / a^{1/2},$$

where β is a numerical constant.

The value of K_c is, however, not a material constant, but depends on various factors including specimen thickness, temperature, and crack velocity. The problems involved in applying the fracture–toughness concept to rate–sensitive materials have been discussed in a review by Kenny and Campbell [105] in which it was pointed out that progress depended on the development of an adequate description of the stress and strain distribution within the plastic zone at the crack tip, such a description incorporating strain–rate sensitivity effects.

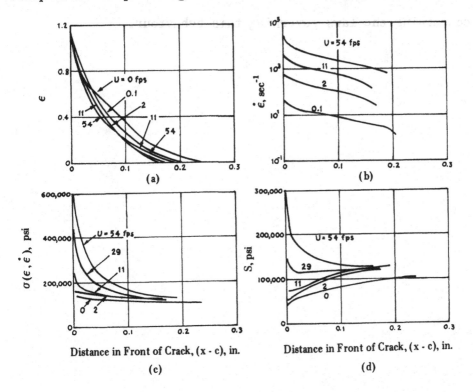

Fig. 65.   Calculations of the influence of crack speed on the plastic zone generated by a propagating ductile crack in the steel foil: (a) Plastic strain gradient, (b) Plastic strain-rate gradient. (c) Flow stress gradient, and (d) Internal stress gradient. The calculations employed N = 32 intervals within the zone , each about 0.007 in wide. [106]

Kanninen et al. [106] have put forward a description of this
kind based on the constitutive relation (1.1.9). They made measurements of the speed of crack propagation in steel-foil specimens, and showed that strain rates exceeding $10^4 s^{-1}$ existed in
the plastic zone (Fig.65, see p.81). It is clear from this and
from later work [107] that the phenomenon of brittle fracture
in mild steel is critically dependent on the high strain-rate
behaviour of the material, and that a full analysis of the problem will only become possible when further information is available concerning the laws governing this behaviour.

References.

[1] V.V. Sokolovsky : Prikl. Mat. Mekh. 12, 261 (1948).

[2] L.E. Malvern : J. Appl. Mech. 18, 203 (1951).

[3] P. Perzyna : Advances in Applied Mechanics Vol. 9, p. 243. (Academic Press, New York, 1966).

[4] U.S. Lindholm : Mechanical Behavior of Materials under Dynamic Loads, p. 77 (Springer, New York, 1968).

[5] K.J. Marsh and J.D. Campbell : J. Mech. Phys. Solids 11, 49 (1963).

[6] T.L. Briggs and J.D. Campbell : Unpublished work.

[7] J. Klepaczko : Arch. Mech. Stos. 19, 211 (1967).

[8] J.D. Campbell : J. Mech. Phys. Solids 15, 359 (1967).

[9] J.D. Campbell and R.H. Cooper : Proc. Conf. on the Physical Basis of Yield and Fracture, p. 77 (Inst. Phys. and Phys. Soc., London, 1966).

[10] J. Klepaczko : Int. J. Mech. Sci. 10, 297 (1968).

[11] W.G. Johnson and J.J. Gilman : J. Appl. Phys. 30, 129 (1959).

[12] J.D. Campbell, J.A. Simmons and J.E. Dorn : J. Appl. Mech. 28, 447 (1961).

[13] J.W. Edington : Mechanical Behavior of Materials under Dynamic Loads, p. 191 (Springer, New York, 1968).

[14] A. Seeger : Phil. Mag. 46, 1194 (1955).

[15] J.D. Campbell and W.G. Ferguson : Ibid., 21, 63 (1970).

[16] T.L. Briggs and J.D. Campbell : Oxford University Engineer-
                                    ing Laboratory Report No. 1091,69 (1969)

[17] F.R.N. Nabarro :   Theory of Crystal Dislocations (Claren-
                                    don Press, Oxford, 1967).

[18] W.P. Mason :   J. Acoust. Soc. Am. 32, 458 (1960).

[19] J.J. Gilman :   Phys. Rev. 20, 157 (1968).

[20] W.G. Ferguson, F.E. Hauser and J.E. Dorn : Br. J. Appl. Phys.
                                    18, 411 (1967).

[21] W.G. Ferguson, A. Kumar and J.E. Dorn : J. Appl. Phys. 38,
                                    1863 (1967).

[22] J. Weertman :   Response of Metals to High Velocity De-
                                    formation, p. 205, (Interscience, New
                                    York, 1961).

[23] H.G. Hopkins :   Engineering Plasticity, p. 277, (Uni-
                                    versity Press, Cambridge, 1968).

[24] J.F. Bell :   J. Appl. Phys. 31, 277 (1960).

[25] L.E. Malvern :   Behavior of Materials under Dynamic
                                    Loading, p. 81 (A.S.M.E., New York,1965)

[26] E.A. Ripperger and H. Watson : Mechanical Behavior of Mate-
                                    rials under Dynamic Loads, p. 294
                                    (Springer, New York, 1968).

[27] Kh. A. Rakhmatulin : Prikl. Mat. Mekh. 12, 39 (1948).

[28] L.E. Malvern :   Quart. J. Appl. Math. 8, 405 (1951).

[29] N. Cristescu :   Dynamic Plasticity (North-Holland, Am-
                                    sterdam, 1967).

[30] J.F. Bell :   Tech. Rep. No. 5, Dept. Mech. Engng.,
                                    The Johns Hopkins University, Baltimore
                                    (1951).

[31] E.J. Sternglass and D.A. Stuart : J.Appl.Mech.20,427 (1953).

[32] B.M. Malyshev :   J. Prikl. Mech. Tech. Fiz. 2, 104 (1961).

[33] J.F. Bell and A. Stein : J. Mécan. 1, 395 (1962).

[34] G. Bianchi :   Stress Waves in Anelastic Solids, p. 101
                         (Springer, Berlin, 1964).

[35] J.W. Craggs :   J. Mech. Phys. Solids 5, 115 (1957).

[36] G.P. DeVault :   Ibid., 13, 55 (1965).

[37] E. Convery and H.Ll.D. Pugh : J. Mech. Engng. Sci. 10, 153
                                                          (1968).

[38] C.H. Yew and H.A. Richardson : Exp. Mech. 9, 366 (1969).

[39] J.D. Campbell and A.R. Dowling : J. Mech. Phys. Solids 18,
                                                       43 (1970).

[40] G.E. Duvall :   Response of Metals to High Velocity Defor-
                         mation, p. 165 (Interscience, New York,
                                                          1961).

[41] J.W. Taylor and M.H. Rice : J. Appl. Phys. 34, 364 (1963).

[42] J.W. Taylor :   Ibid., 36, 3146 (1965).

[43] R.H. Cooper and J.D. Campbell : J. Mech. Engng. Sci. 9, 278
                                                          (1967).

[44] D.S. Clark and D.S. Wood : Proc. Am. Soc. Test. Mat. 49, 717
                                                          (1949).

[45] A.H. Cottrell :   Properties of Materials at High Rates of
                         Strain, p. 1 (Inst. Mech. Engrs., London,
                                                          1957).

[46] A.H. Cottrell and B.A. Bilby : Proc. Phys. Soc. A62, 49
                                                          (1949).

[47] J.D. Campbell :   Acta Met. 1, 706 (1953).

[48] G.T. Hahn :   Ibid., 10, 727 (1962).

[49] A.H. Cottrell :   The Relation between the Structure and
                             Mechanical Properties of Metals, p.455
                             (H.M.S.O., London, 1963);

[50] J.D. Campbell, R.H. Cooper and T.J. Fischhof: Dislocation
                             Dynamics, p.723 (McGraw–Hill, New york,
                             1968.

[51] J.D. Campbell and K.J. Marsh : Phil. Mag. $\underline{7}$, 933 (1962).

[52] T. Yokobori :   Phys. Rev. $\underline{88}$, 1423 (1952).

[53] R.J. Aspden and J.D. Campbell : Proc. Roy. Soc. A $\underline{290}$, 266
                                                              (1966).

[54] U.S. Lindholm :   Behavior of Materials under Dynamic
                            Loading, p. 42 (A.S.M.E., New York,
                                                              1965).

[55] J. Klepaczko :   Int. J. Solids Structures $\underline{5}$, 533 (1969).

[56] J. Klepaczko :   J. Mech. Phys. Solids $\underline{16}$, 255 (1968).

[57] T.L. Briggs: Unpublished work.

[58] H. Kolsky :   Proc. Phys. Soc. B $\underline{62}$, 676 (1949).

[59] J.M. Krafft, A.M. Sullivan and C.F. Tipper : Proc. Roy.
                            Soc. A $\underline{221}$, 114 (1954).

[60] J.D. Campbell and J. Duby : Ibid., A $\underline{236}$, 24 (1956).

[61] F.E. Hauser, J.A. Simmons and J.E. Dorn : Response of Me-
                            tals to High Velocity Deformation, p.93
                            (Interscience, New York, 1961).

[62] L. Efron and L.E. Malvern : Exp. Mech. $\underline{9}$, 255 (1969).

[63] E.D.H. Davies and S.C. Hunter : J. Mech. Phys. Solids $\underline{11}$,
                                                              155 (1963).

[64] U.S. Lindholm :   Ibid., $\underline{12}$, 317 (1964).

[65] S. Yoshida and N. Nagata : Trans. Jap. Inst. Met. $\underline{7}$, 273
                                                              (1966).

[66] C.J. Maiden and S.J. Green : J. Appl. Mech. 33, 496 (1966).

[67] J. Harding, E.O. Wood and J.D. Campbell : J. Mech. Engng.
        Sci. 2, 88 (1960).

[68] J.D. Campbell and J. Harding : Response of Metals to High
        Velocity Deformation, p. 51 (Interscience,
        New York, 1961).

[69] J. Harding :   High Energy Rate Working of Metals, p. 365
        (Norway, 1964).

[70] W.E. Baker and C.H. Yew : J. Appl. Mech. 33, 917 (1966).

[71] J. Duffy, J.D. Campbell and R.H. Hawley : Ibid.
        (to be published).

[72] A.R. Dowling, J. Harding and J.D. Campbell : J. Inst. Met.
        (in press).

[73] W.F. Greenman, T. Vreeland and D.S. Wood : J. Appl. Phys.
        38, 3595 (1967).

[74] J.A. Gorman, D.S. Wood and T.Vreeland : Ibid., 40, 833
                                                          (1969).

[75] J. Harding :   J. Mech. Engng. Sci. 7, 163 (1965).

[76] J. Harding :   Unpublished work.

[77] A.G. Atkins :   Oxford University Engineering Laboratory
        Report No. 1056, 68 (Also J. Inst. Met. 97,
        289 (1969).

[78] S.J. Green and C.J. Maiden : Unpublished work.

[79] C.H. Karnes and E.A. Ripperger : J. Mech. Phys. Solids 14,
                                                          75 (1966).

[80] A.G. Atkins and D.P. Porter : U.S. Steel Corp. Applied Re-
        search Laboratory Report (1967)

[81] M.G. Stevenson and P.L.B. Oxley : Proc. Inst. Mech. Engrs.
                                                          184 (1970).

[82] P.L.B. Oxley and M.G. Stevenson : J. Inst. Met. 95, 308 (1967).

[83] R.G. Fenton and P.L.B. Oxley : Proc. Inst. Mech. Engrs. 184, (1970).

[84] C. Zener and J.H. Holloman : J. Appl. Phys. 15, 22 (1944).

[85] W. Johnson and R.A.C. Slater : Proc. Inst. Mech. Engrs. 179, 257 (1964).

[86] W. Johnson and F.W. Travis : Ibid., 180, 197 (1966).

[87] R. Davies and S.M. Dhawan : Ibid., 180, 182 (1966).

[88] A.R. Dowling and J. Harding : First. Int. Conf. of the Center for High Energy Forming, vol. 2, p. 7.3.1 (Denver, 1967).

[89] — High Energy Rate Working of Metals (Nato Advanced Study Institute, Sandefjord – Lillehammer, Norway, 1964).

[90] — First Int. Conf. of the Center for High Energy Forming (Denver, Colorado, 1967).

[91] — Second Int. Conf. of the Center for High Energy Forming (Denver, Colorado, 1969).

[92] B. Rawlings : Proc. Roy. Soc. A 275, 528 (1963).

[93] E.W. Parkes : Ibid., A 228, 462 (1955).

[94] S.R. Bodner and P.S. Symonds : Plasticity, p. 488 (Pergamon, New York, 1960).

[95] S.R. Bodner and P.S. Symonds : J. Appl. Mech. 29, 719 (1962).

[96] B. Rawlings : J. Mech. Engng. Sci. 6, 327 (1964).

[97] B. Rawlings : Proc. Inst. Civ. Engrs. 29, 389 (1964).

[98] B. Rawlings : Int. J. Mech. Sci. 9, 633 (1967).

[99] J.M. Kelly : Int.J. Solids Structures 3, 521 (1967).

[100] J.M. Kelly and T.R. Wilshaw : Proc. Roy. Soc. A 306, 435
                                                                      (1968).

[101] S.R. Bodner :   Engineering Plasticity, p. 77 (University
                         Press, Cambridge, 1968).

[102] P.S. Symonds :   Ibid., p. 647.

[103] A.A. Griffith : Phil. Trans. Roy. Soc. A 221, 163 (1920).

[104] G.R. Irwin :   Encyclopaedia of Physics, Vol. 6, p. 551
                         (Springer, Berlin, 1958).

[105] P. Kenny and J.D. Campbell : Progress in Materials Science,
                         vol. 13, p. 135 (Pergamon, Oxford, 1968).

[106] M.F. Kanninen, A.K. Mukherjee, A.R. Rosenfield and
       G.T. Hahn :   Mechanical Behavior of Materials under Dy-
                         namic Loads, p. 96 (Springer, New York,
                         1968).

[107] G.T. Hahn, M.F. Kanninen and A.R. Rosenfield : Fracture
                         (Proc. Second. Int. Conf. on Fracture)
                         p. 58 (Chapman and Hall, London, 1969).

Contents.

Printed in the United States
By Bookmasters